A Matter of Change

Harcourt

SCHOOL PUBLISHERS

Orlando   Austin   New York   San Diego   Toronto   London

Visit *The Learning Site!*
www.harcourtschool.com

Mix It Up!

You make a **mixture** when you mix together kinds of matter. Fruit salad is a mixture of solids.

Substances in a mixture do not become other substances. In fruit salad, pieces of apples are still apples. Pieces of oranges are still oranges.

These pieces of fruit do not change when they are mixed.

Mixtures can have pieces of things in different sizes and amounts.

You can make a mixture from two liquids. Chocolate syrup and milk make chocolate milk. You can also make a mixture from different gases.

Mixtures can mix matter from different states, too. Salt water is a mixture of a liquid and a solid—water and salt. Bubbly drinks mix liquids and gases.

CAUSE AND EFFECT What happens to oranges when they become part of fruit salad?

Three Forms of Water

Water is the only matter that is naturally found in three forms—solid, liquid, or gas.

Water changes from a liquid to a solid when enough heat is taken away. Think about putting water into the freezer. When water becomes very cold, it changes to ice!

Can you see two forms of water in the glass?

Ice is water in its solid form. When water is a solid, it has its own shape.

Ice melts when its temperature is high enough. Think about an ice cube in the sun. It turns from a solid to a liquid. It does not have its own shape anymore.

Focus Skill

CAUSE AND EFFECT How does the solid form of water change when enough heat is added?

As a solid, water has a shape. As a liquid, it does not.

Evaporation

Water can be a gas. The liquid can turn to a gas when enough heat is added to it. This change from liquid to gas is called **evaporation**.

The sun's heat makes water from the ocean, lakes, and other bodies of water evaporate.

On a hot day, some of this water evaporates.

You can see evaporation happen. Have you ever seen water boiling in a pot? The water seems to go away. The water is turning into a gas that mixes with the air. You are watching evaporation.

The water that evaporates becomes water vapor. It is in the air. You cannot see it.

When water is a gas, it is called **water vapor**.

 CAUSE AND EFFECT What happens to boiling water so that you no longer see it?

Condensation

Water can change from a gas back to a liquid. This change is called **condensation**. When enough heat is taken away from water vapor, it changes to water.

Condensation happens when a gas cools and changes to a liquid.

You do not have to be a scientist to watch condensation happen!

You can see condensation happen. Have you ever seen tiny drops of water on the outside of a cold glass? The cold glass takes heat from the air around it. This changes the water vapor into a liquid.

The tiny drops of water on the outside of the glass were water vapor only seconds ago.

CAUSE AND EFFECT What happens to water vapor when it gets very cold?

Changing Properties

Properties of things can change when they are heated or cooled. Water can be a solid, a liquid, or a gas. These states depend on water's temperature. Water can even change so we can not see it when it becomes a gas.

 CAUSE AND EFFECT An ice cube is taken from the freezer and placed outdoors. It does not melt. Why?

If this hot water is made very cold, it will become ice.

Fast Fact

A drop of water may travel thousands of miles between the time it evaporates and the time it falls to Earth again.

All of this water was once a gas called water vapor.

Summary

You can put different kinds of matter together to make mixtures. Water can be a solid, a liquid, or a gas. Water's properties depend on how much heat is taken from it or added to it.

Glossary

condensation The change of water from a gas to a liquid. Condensation happens when heat is taken away from water vapor. (8, 9)

evaporation The change of water from a liquid to a gas. Evaporation happens when heat is added to liquid water. (6, 7)

mixture A mix of different kinds of matter. Substances in a mixture do not become other substances when they are mixed. (2, 3, 11)

water vapor Water that is in the form of a gas (7, 8, 9, 11)